Orlando Austin New York San Diego Toronto London

Visit *The Learning Site!*
www.harcourtschool.com

Lesson 1

How Do Forces Affect Us?

VOCABULARY

velocity
force
acceleration
inertia

Velocity tells us both the speed of a moving object and its direction. Look for the velocity of the jet ski and the geese.

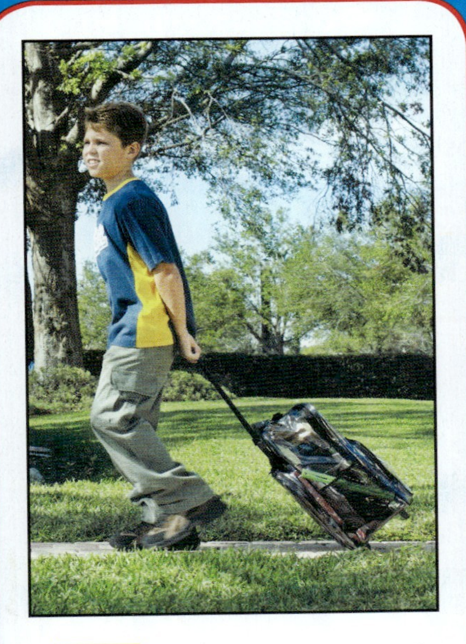

A **force** is a push or a pull. This boy is using force to pull the backpack.

Acceleration is an object's change in velocity divided by the time it takes for the change to occur. This car is accelerating.

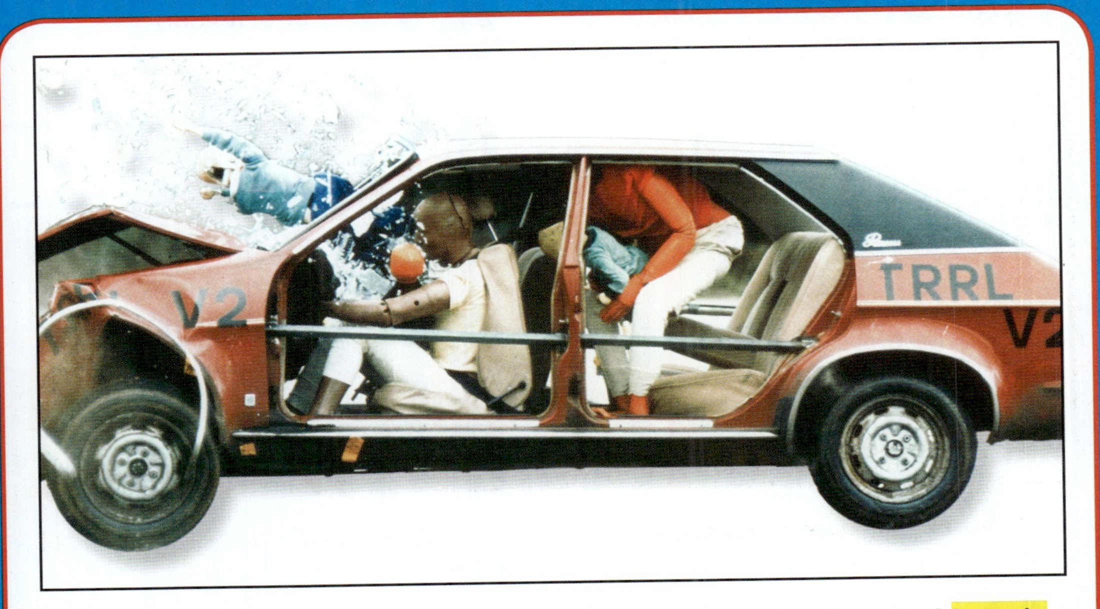

The tendency of matter to resist a change in its motion is called **inertia**. The dummies keep going forward because of inertia.

READING FOCUS SKILL

MAIN IDEA AND DETAILS

The **main idea** is what the text is mostly about. **Details** are pieces of information about the **main idea**.

Look for information about how forces affect objects and **details** about the factors that affect each force.

Describing Motion

Think of some words that you use to describe motion. Turning, sliding, and twisting are all types of motion. There are many kinds of motion. But all types of motion have one thing in common. Motion is change in position.

How can you tell something is moving? You use a *frame of reference*. This is a stationary object or group of objects you use to see if you're moving. For example, if you ride your bike past a building, you know you're moving. Your position changes compared with the building. The building is stationary. It is not moving. The building is your frame of reference.

Velocity

The picture shows that direction is often expressed in terms of the compass.

The lifeguard chair is a frame of reference for the swimmer.

The swimmer is moving west at a steady pace.

One way to describe motion is to tell what its speed is. *Speed* is the distance something moves in a certain amount of time. Suppose you ride 150 kilometers (93 miles) in 2 hours. To find your speed, divide the distance traveled by the time. Your speed was 150 kilometers ÷ 2 hours = 75 kilometers per hour.

Velocity tells both the speed of an object and its direction. For example, you would have to add your direction to the speed above to give your velocity. You might say your velocity is 75 kilometers per hour north.

The picture shows the velocities of some things at the beach. Notice that each velocity tells both the speed and direction. The birds are moving in one direction at the same speed. This means that they have a constant velocity. If the birds start flying in a circle, their velocity will change. Velocity changes any time speed or direction changes.

We often use arrows to show velocity. The length of the arrow shows the speed measurement. The direction it points shows the direction measurement.

What is your speed if you ride 38 kilometers in 2 hours on your bike?

What Forces Do

A **force** is a push or a pull. If you pull on a door it opens. If you push on the door it closes. The force to open and close the door comes from your muscles. There are other forces, too. Gravity and magnetism are both forces.

Forces slow things down, speed things up, stop them, start them, and turn them. A force is at work whenever an object changes velocity. A change of velocity is called **acceleration**. Acceleration is an object's change in velocity divided by the time it takes for that change to occur.

Let's look at an example. A train sitting in a station has a velocity of 0 km/hr north. As it leaves the station, it reaches a velocity of 75 km/hr north in 5 seconds. You subtract to find the change in velocity. The change in velocity is 75 km/hr minus 0 km/hr in a northerly direction.

To find the acceleration, you need to divide. Divide the velocity change you just found by the time it took for the change to happen. 75 km/hr ÷ 5 seconds = 15 km/hr per second. So, the train's rate of acceleration was 15 km/hr per sec. This means that for each second the train traveled, its speed increased by 15 km/hr.

As the train reaches the next station, its speed decreases. The velocity of the train changes. So the train is accelerating. Acceleration occurs any time velocity changes, even if something is slowing down.

◀ **Acceleration occurs both when speed increases and when speed decreases.**

Isaac Newton (1642–1727) was one of the world's greatest scientists. He developed laws describing forces. One law tells how force relates to mass and acceleration.

This law says that the greater the force on an object, the greater its acceleration. Think of opening a door. If you tap the door, it barely moves. But if you shove the door with more force, it swings wide open. It accelerates faster.

The law also says that if the same force acts on two objects, the object with less mass accelerates more than the object with more mass. Think of pushing a shopping cart. If you push an empty cart with the same force as you push a full cart, the empty cart will accelerate more.

What happens to the amount of force needed to move an object when its mass increases?

◀ The girl kicks the ball gently. She doesn't use a lot of force. The ball has little acceleration.

▼ This girl kicks the ball with more force. The ball has greater acceleration.

Inertia

For many years, people thought that an object would stop moving when you removed the force that put it in motion. For example, a marble rolling on the floor stops rolling when you stop pushing it.

Isaac Newton and other scientists showed that this idea was wrong. Motion itself does not require a force. It is a change in motion that requires a force. Think of the marble again. The marble stops rolling because the force of friction stops it. If there were no friction, the marble would keep rolling.

An object will not change its motion unless a force acts on it. It will not slow down, speed up, stop, start, or turn without a force acting on it. This tendency of matter to resist change in its state of motion is called **inertia**.

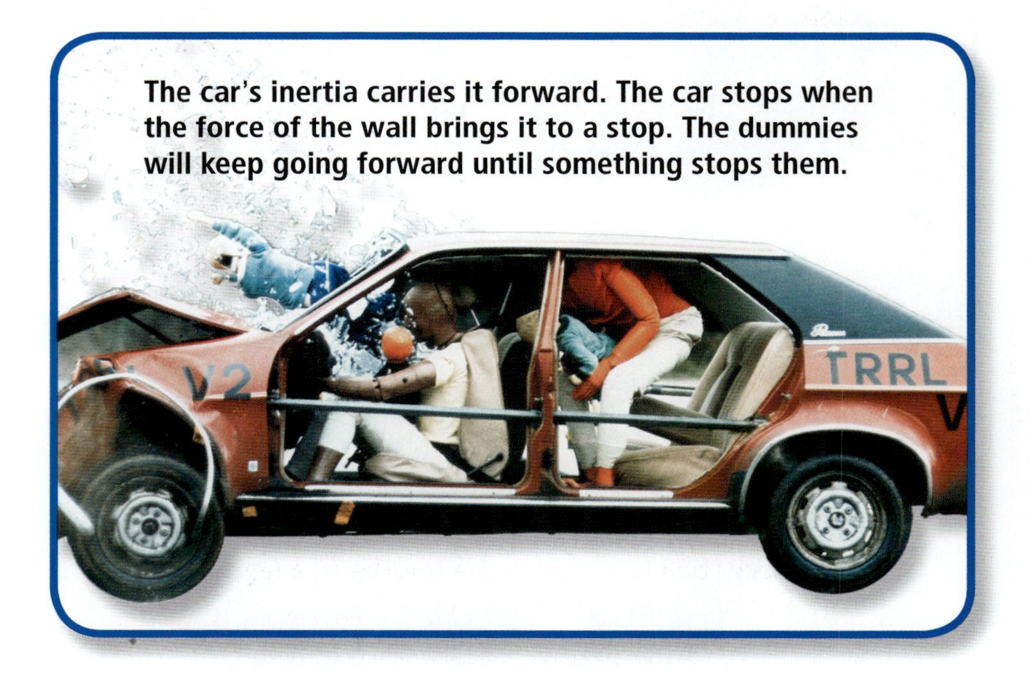

The car's inertia carries it forward. The car stops when the force of the wall brings it to a stop. The dummies will keep going forward until something stops them.

One of Newton's laws is called the *law of inertia*. This law says that an object at rest stays at rest and an object in motion stays in motion at a constant velocity, unless the object is acted upon by an outside force.

In other words, an object will not accelerate without a force acting on it. Force must overcome an object's inertia to make it move.

What is inertia?

The boy must use force to overcome the backpack's inertia.

Review

Focus Skill

Complete this main idea statement.

1. Any push or pull is a ______.

Complete these detail statements.

2. If you push two objects with the same force, the object with more ______ accelerates less.
3. A soccer ball is sitting in the middle of a field and does not move. This is an example of ______.
4. The change in an object's velocity is called ______.

Lesson 2

How Do Forces Interact?

VOCABULARY

balanced forces
unbalanced forces
friction

Equal forces that act in opposite directions on an object are called **balanced forces**. Gravity pulls the vase downward. The support force of the table pushes the vase upward. The forces cancel each other. Because the forces are balanced, the vase does not move.

Forces that do not cancel one another are called **unbalanced forces**. An object with unbalanced forces acting on it will accelerate.

Friction is a force that acts between any surfaces in contact with each other. Friction prevents motion or slows it down.

READING FOCUS SKILL

COMPARE AND CONTRAST

You **compare and contrast** when you look for ways things are similar and different. **Compare** means to find the way things are similar. **Contrast** is to find the way things are different.

Compare and contrast balanced and unbalanced forces.

Balanced and Unbalanced Forces

Equal forces acting in opposite directions on an object are called **balanced forces**. You may have seen balanced forces during a tug-of-war. When both teams pull the rope with equal forces, the rope does not move.

Balanced forces cancel one another. They have the same result as no force at all. An object affected by balanced forces will not accelerate. It will not start, stop, speed up, slow down, or turn. It will continue at a constant velocity.

The total of all the forces acting on an object is called the *net force*. The net force is zero when forces are balanced.

The force of gravity pulls the vase downward. The support force of the table pushes the vase upward. The forces acting on the vase are balanced. What is the net force?

Usually more than one force acts on an object. Sometimes these forces are not balanced. Forces that do not cancel one another are called **unbalanced forces**. An object with unbalanced forces acting on it will accelerate. Its velocity will change. With unbalanced forces, the net force is not zero.

You use unbalanced forces every time you move. Unbalanced forces help you do things like push your skateboard or throw a ball.

Suppose you have to move boxes. You go to one box, apply a force, and move the box to the right. Then your friend comes. Your friend gets on the other side of a box. You both push the box with the same force, but in opposite directions. The box will not move. The forces are balanced. But suppose your friend gets on the same side of the box you're on. You both push the box in the same direction. Then the box will move with the force you are using plus the force your friend uses.

What is the net force of an object with unbalanced forces acting on it?

▲ **The forces acting on the sleigh are unbalanced as it turns the corner.**

Forces in Our World

Can you think of all the ways you use forces? You use your muscles in many ways during the day to apply forces.

Many forces do not come from human muscles. Look at the water strider in the picture below. This insect is always being pulled down by gravity, but it is not accelerating. So you know that another force is acting on the insect. What force balances gravity? *Surface tension* acts in the opposite direction of gravity and with equal strength. This force occurs on the surface of a liquid. Surface tension pulls the particles at a liquid's surface together to make the smallest possible surface area. Different liquids have different surface tensions. Water has a greater surface tension than gasoline or alcohol.

Things float on water because of the *buoyant force*. The buoyant force pushes upward on objects in a liquid. It works against gravity. When gravity and the buoyant force balance, an object floats. When gravity is greater, the object sinks.

Magnetic force is a push or pull that works between magnetic poles. The magnetic force pulls opposite poles (North and South) together. It also pushes like poles (North and North or South and South) apart.

Gravity pulls the parachutist downward. A force called *air resistance* pushes upward. Air resistance acts to slow the parachutist, and he floats to Earth at a constant velocity.

Forces even act inside tiny particles of matter. These forces are called *nuclear forces*. Nuclear forces hold atoms together.

Electric force acts between objects that have electric charge. There are positive electric charges and negative charges. The electric force pulls objects with different charges together. A positive charge attracts a negative charge. Electric force pushes objects with the same charge apart. Two negative charges repel each other. Two positive charges also repel each other.

Have you ever brushed your hair and watched your hair move toward the brush? This happens when your hair and the brush have opposite electric charges. The electric force pulls your hair and the brush together.

Gravity, surface tension, the electic force, magnetic force, the buoyant force, and others can all act at the same time. Even with many forces acting at once, the forces may still be balanced. So the object maintains its state of motion. If the forces are unbalanced, the object will change its velocity.

Describe how buoyant force allows a ball to float in a pool.

Friction Opposes Motion

Friction is a force that acts between any surfaces in contact with each other. It either prevents motion or slows it down. Friction is part of daily life. When you use an eraser, there is friction between the eraser and the paper. Friction makes it hard to move the eraser easily. There is also friction between your feet and the ground. When you walk on a rough surface, like grass, there is more friction than when you walk on a smooth surface, like a tile floor.

Sometimes people need to increase friction. When surfaces are very slippery, someone could fall. People put sand on icy sidewalks to provide more friction. Some people wear shoes with rubber soles to increase friction. The friction helps keep people from slipping and falling on slippery surfaces.

The air hockey puck rides on a thin layer of air. This reduces friction to almost zero.

Sometimes people want to lessen the amount of friction between two surfaces. People put grease or oil on machine parts to lubricate them. This helps smooth away bumps. The machine parts can now slide by one another more easily. When machines are lubricated, the parts do not produce as much heat. This helps the machine last longer.

When is it helpful to have a lot of friction when walking?

This girl has to apply enough force to overcome friction.

Review

Focus Skill

Complete these compare and contrast statements.

1. ______ ______ cancel one another out and ______ ______ do not cancel one another out.
2. In both ______ force and ______ force opposite forces attract and like forces repel.
3. Increasing ______ helps make slippery surfaces safer, but can damage machine parts.

What Is Gravitational Force?

VOCABULARY

gravitational force
weight

Gravitational force is a force of nature. It acts between any two masses in the universe. It pulls them toward one another.

Weight is the measurement of the force of gravity on an object. In the microgravity of space, astronauts lose their sensation of weight.

READING FOCUS SKILL

CAUSE AND EFFECT

A **cause** is what makes something happen. An **effect** is what happens.

Look for how gravitational force **causes** an **effect** on matter.

Gravitational Force

Gravitational force is a basic force of nature. It works all the time. It acts everywhere. **Gravitational force** acts between any two masses in the universe. It pulls them toward one another.

You are already familiar with Earth's gravitational force. It is what pulls objects, like balls and raindrops, toward Earth's surface. But gravitational force acts between any two masses.

The diver uses her muscles to push herself into the air. Then gravity brings her down into the water. ▼

All masses attract one another. Right now, you are being pulled toward the bookshelf in your classroom. You are also being pulled by the trees outside, your bed at home, and even the moon—all while sitting at your desk at school! You don't feel these pulls because the gravitational force is so small. At the same time, your mass is also pulling all of these objects toward you.

Gravity pulls this boy down the hill.

We are most aware of the gravitational pull of Earth on objects at Earth's surface. This force pulls us and everything around us toward the center of Earth. This force of gravitation is also called *gravity*.

Gravity works in many ways. It pulls balls and raindrops to Earth's surface. It also helps keep oceans on the surface of Earth. It keeps the water surface fairly level. Gravity keeps our atmosphere from flying out into space.

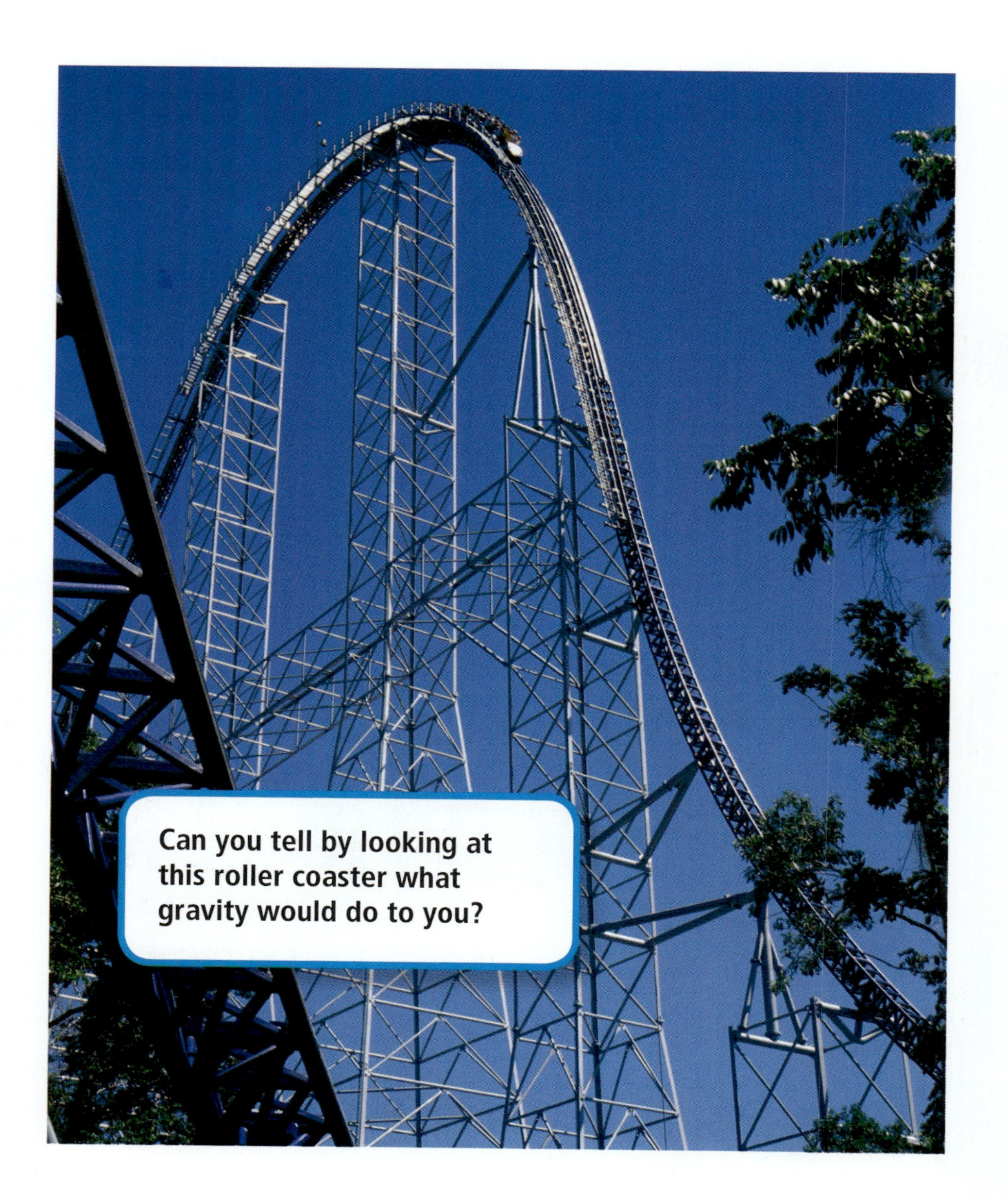

Can you tell by looking at this roller coaster what gravity would do to you?

Gravity is different from forces people use to push and pull. When you push or pull something, you need to contact it. For example, your hand needs to touch the door to push it open. Gravity is an "action-at-a-distance" force. Gravity acts through empty space. Gravity acts without contact.

What effects does gravity have on Earth's oceans?

All masses are pulled toward one another through gravitational force. Even this boy and his dog are pulled toward each other ever so slightly because of this force.

How Mass and Distance Affect Gravitation

Suppose you put two tennis balls on a table. Gravitational force acts to attract them to each other. But they don't move toward each other. That's because the gravitational force between them is very weak. But take one of the tennis balls and hold it above the floor. Let go of it. You'll see it drop to the floor. The gravitational force between the ball and Earth is very strong. The ball moves toward the center of Earth.

The strength of gravitational force depends on two things. It depends on the masses of the objects and it depends on the distance between them. The more massive two objects are, the stronger the gravitational force. The closer they are, the stronger the gravitational force.

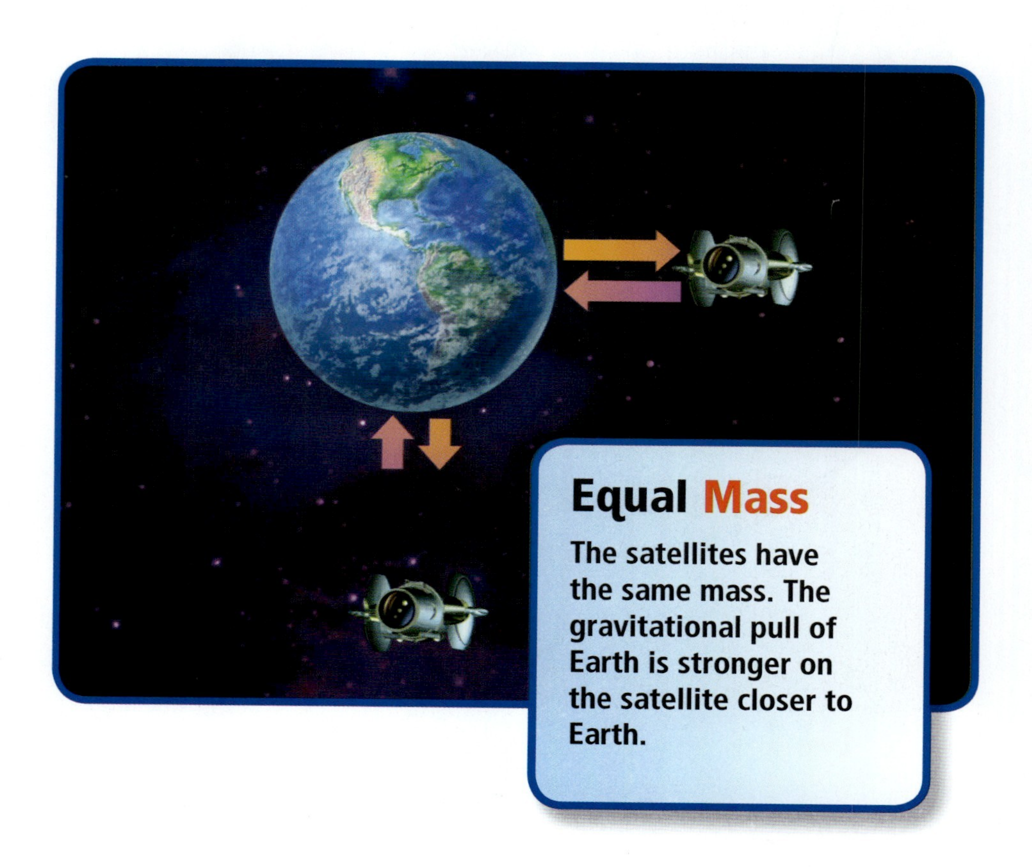

Equal Mass

The satellites have the same mass. The gravitational pull of Earth is stronger on the satellite closer to Earth.

The two tennis balls have small masses. So the force of attraction between them is weak. They don't move. Earth has a very large mass. The ball you hold is close to Earth. So the ball moves.

The sun also has a gravitational pull on you. But it is so far away you don't notice it. We live on Earth's surface. So we notice the pull of Earth's gravity.

Gravitation is important in the solar system. Because the masses of the planets are so huge, all of the planets attract one another. The sun has a greater mass than any of the planets. So its pull is greater. This pull makes all of the planets go around the sun.

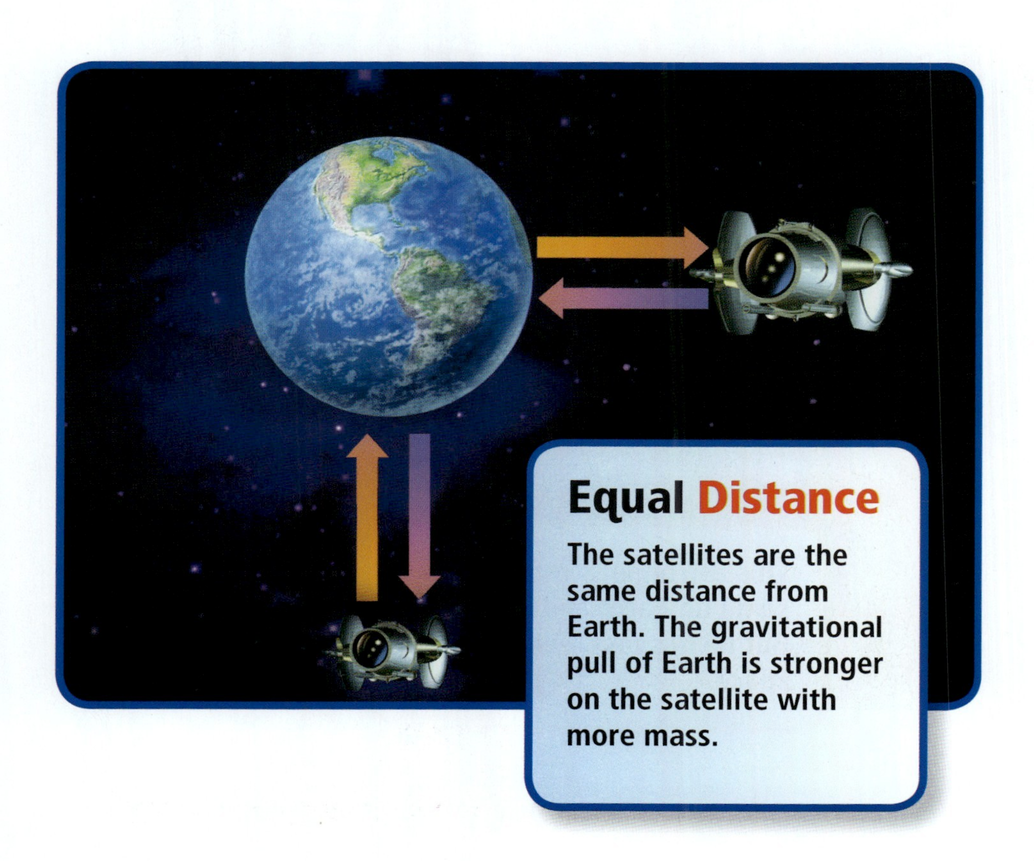

Equal Distance

The satellites are the same distance from Earth. The gravitational pull of Earth is stronger on the satellite with more mass.

Why don't the planets just fall into the sun? They follow nearly circular paths, or orbits, around the sun. This is because of gravitational force and inertia. Remember that objects travel in straight lines and at constant speeds unless they are acted on by net forces. If no net forces acted on a planet, it would travel in a straight line forever because of its inertia. But the gravitational force between the planet and the sun combines with the planet's inertia. This causes the planet to travel in a curved path around the sun instead of in a straight line into space.

Why do planets move in curved orbits instead of straight lines?

This astronaut's mass is the same on Earth and in space. But he has less weight in space.

Gravity and Weight

The force of gravity on any object depends on an object's mass. The greater the mass, the greater the force of gravity acting on it. **Weight** is the measurement of the force of gravity on an object. Weight and mass are not the same.

Mass is the amount of matter in an object. The mass stays the same no matter where the object is located. Mass is measured using a balance.

Weight varies from place to place depending on gravitational forces. A person weighs less on the moon than on Earth because the force of gravity is weaker on the moon. Weight is measured using a scale.

How does gravitational force affect weight?

Review

Focus Skill

Complete these cause and effect statements.

1. ______ ______ acts between any two masses and pulls them toward one another.
2. The huge mass of the sun causes all of the planets to travel in ______ around the sun.
3. If two objects have the same mass, the object closer to Earth will experience a ______ gravitational force.
4. The ______ of a person is less on the moon than on Earth because of gravitational force.

GLOSSARY

acceleration (ak•sel•er•AY•shuhn) A change in velocity divided by the time it takes for that change to occur.

balanced forces (BAL•uhnst FAWRS•iz) Equal forces that act in opposite directions and cancel one another out.

force (FAWRS) A push or a pull.

friction (FRIK•shuhn) A force that acts between any two surfaces in contact with one another by preventing or slowing motion.

gravitational force (grav•ih•TAY•shuhn•uhl FAWRS) The force that acts between any two masses in the universe and pulls them toward one another.

inertia (in•ER•shuh) The tendency of matter to resist a change in its state of motion.

unbalanced forces (uhn•BAL•uhnst FAWRS•iz) Forces that do not cancel one another out.

velocity (vuh•LAHS•uh•tee) The speed and direction of a moving object.

weight (WAYT) The measurement of the force of gravity on an object.